Getting Ahead in Science

Terry Cash

Longman

Longman Group UK Limited

Longman House
Burnt Mill, Harlow, Essex CM20 2JE, England
and Associated Companies throughout the World

First published 1990
ISBN 0 582 05868 6

British Library Cataloguing in Publication Data

Cash, Terry
    Getting ahead in science.
    1. Science
    I. Title II. Series
    500

    ISBN 0-582-05868-6

Set in 11/12pt Frutiger Light and Frutiger Bold
on Scantext 1000 system.

Produced by Longman Group (FE) Ltd.
Printed in Hong Kong

Contents

Contents

Introduction

Science is now an important core subject in every school's curriculum and an understanding of science is so important for our children who will work in a world that is constantly changing and which is becoming more and more dependent upon scientific knowledge and progress.

Tomorrow's world will belong to the Chemist, the Biologist, the Physicist and the Mathematician. This book is for the concerned parent who would like her child to get ahead in her academic career and play a valuable role in the world of the future.

With step by step instructions it guides the reader through simple but important activities that lay the foundation for a clear understanding of scientific processes and principles, and, perhaps what is more important, it makes learning fun.

Great emphasis is placed upon learning by doing things. No previous experience of science is required or expected, just a willingness to try activities chosen from over eighty experiments and investigations offered.

Many of these activities begin with important information which explains, clearly and simply, the scientific principles involved and the scientific terminology used. There is also a comprehensive list of items required for your child to tackle each task. However it is fully appreciated that expensive, specialized equipment is not easily obtainable which is why most of the experiments and investigations can be done with nothing more than items of household junk. In fact the only thing that is needed is the sense of curiosity and inquisitiveness that every child has as a natural gift . . . happy experimenting!

The living world

The human body

Our bodies are amazing natural machines. The many different parts do a wide range of jobs with little or no outside maintenance for years and years. For example our heart is a special muscle that pumps the blood to the lungs to pick up fresh oxygen and then all round the body. The heart beats roughly 60 times a minute which means that the heart of a 70 year old person will have pumped about two thousand million times without skipping a beat!

Heart beat

A doctor can listen to your heart beats with a stethoscope. You can make your own stethoscope by pushing the end of a small plastic funnel into the end of a short length of plastic tubing. Place the funnel over the left side of your chest and hold the other end to your ear. If it is very quiet you should be able to hear the thump of your own heart beat. Another way to measure your heart beat rate is to rest the first two fingers of your right hand gently on your left wrist between the centre and left side of the wrist.

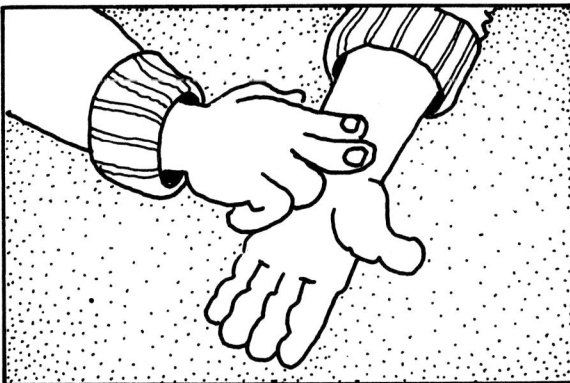

With care you should find the spot where you can feel the clear thump of your pulse. Sit quietly for a while and count your heart or pulse beats for one minute. Now go for a fast run around the garden. Sit down and count your heart beat rate again. What do you notice? Check the heart beat rates of adults and young children, you may find quite a difference. A child's heart beat rate is usually much higher than an adult's. When resting, an adult's heart may only beat 50 times a minute, but a child's could beat 80 times or more. Why do you think the heart beats so much faster when we work hard or run fast?

Plenty of puff

While you check out your heart rate you can check your breathing rate at the same time. Make a note of how often you breathe in a minute while you are resting quietly. After you have run quickly round the garden count the number of breaths you take now. To get more oxygen to the muscles that are suddenly having to work so hard you breathe more often and the heart beats faster to pump the oxygen-rich blood round your body.

Now find out how much puff you have got. All you need is a milk bottle, 3 metres (10 feet) of plastic tubing (the kind sold by large chemists in their home-made wine and beer departments) and some plasticine or modelling clay.

Cut about 1 metre (3 feet) from the end of the tubing and tape the longer length to a suitable door frame. Also tape a strip of paper to the door frame to record results. Half fill the bottle with water, add a few drops of food colouring and wedge the ends of both pieces of tube into the bottle with the plasticine. Make sure that the end of the shorter tube is just inside the neck of the bottle, but the longer tube should be as far into the water as possible.

Take a deep breath and blow as hard as you can down the short tube. This forces the water up the longer tube. As you blow, get someone to mark the highest point that the water is pushed up the tube. Let everyone in the family have a go. Who has the greatest puff? Is anyone strong enough to blow the water out of the top of the tube?

Muscles and bones

A skeleton could not walk about by itself, it needs muscles to make it move. Muscles work by pulling and often come in pairs so that when one is pulling the other is being stretched. As an example, place your left hand on the upper part of your right arm. Now clench your right fist and bring it up towards your shoulder. Can you feel

the muscle pulling and tightening? This muscle is called your biceps. To make it go down again a muscle underneath the top part of your arm, called the triceps, pulls your arm down and at the same time stretches the biceps so that it is ready to work once more.

How strong are you?

Muscles can be very strong. How strong are you? Try these tests on your family and friends to see who is strongest. Take some ordinary bathroom scales in both hands and squeeze them as tightly as you can. Get someone to check with you how far round the dial moves. Make a note of the reading in kilogrammes and check from time to time to see if your grip is getting stronger.

Leg power

Another way of testing your muscles is to ask someone to hold the scales against a suitable wall about half a metre (2 feet) above the ground. Lay on your back and place the soles of your feet on the scales. When you are in a comfortable position, push as hard as you can against the scales. Ask your helper to read off how far the dial moves. Can you push so hard that the reading on the dial is almost as much as your own weight?

How are your nerves?

Nerves carry messages all round our bodies. Most of the messages travel to or from the brain. Some of the messages go to our muscles to make them pull at the right time so that we can smile, clap, walk and talk. Even to be able to read this book needs lots of muscles to be

working together in the right way. Some of our movements we don't even have to think about, they happen by themselves. They are called reflex movements. For instance, stand in front of a mirror in a dark room and look at your pupils, the black holes in the centre of your eyes that let light in. Because it is dark they will be very large to let in as much light as possible. Now flash the beam of a torch into one of your eyes, the pupil immediately closes to shut out the unexpected light. You don't have to think about it, it just happens.

Another example of a reflex is the knee jerk. Sit in a chair with one leg crossed over the other. Now hit your knee just below the bottom of your knee cap with the edge of a hard book. If you hit the right spot your leg will jump. You have no control over this, it just happens. Doctors use special little hammers to tap various parts of the body to check reflexes. It gives them important information about the condition of a patient's nervous system, especially if the person has had a bad fall or has been in an accident.

Thinking distance

Your reaction time is the time that it takes for your nerves to send a message to the brain and for your brain to decide what to do and send messages to the right muscles to make the right movements. In road safety, when stopping in an emergency, this is called the "thinking distance" which is how far your bicycle or a person's car will travel while you are still thinking about putting on the brakes.

How fast are your reactions?

All you need for this investigation is an ordinary school ruler about 30cm long and a friend to help you.

Ask your friend to hold the end of the ruler so that it hangs down. You must place your thumb and first finger either side of the bottom of the ruler. Don't touch the ruler, but be ready to grab it. Ask your friend to let go of the ruler at any time they choose. You must be ready to nip your thumb and finger together to stop the ruler from falling on the floor when she lets go. Start with the 0cm mark between your fingers and see how far the ruler falls before you grab it. The shorter the distance in centimetres, the faster your reactions. Try this one on your family too, to see who is really quick off the mark.

Caring for living things

You can learn so much just by sitting and watching what is going on around you. How does a spider spin her web? Do flowers turn to follow the sun? Can ladybirds fly? Sometimes it is necessary to disturb things a little to spot what is happening. But if you turn over a rock to find an ants' nest or lift a log to watch the snails, always return the things to their original place very carefully. Try to leave places as undisturbed as possible.

There is always a strong temptation to bring things back to the house for a better look, but never collect creatures in this way unless you can look after them properly in a specially made vivarium and even then they should be returned to the wild as soon as possible. In some places the natural population of frogs has almost been wiped out because over enthusiastic nature watchers keep taking all the frog spawn home in jars. The tadpoles die in the jars and none of them have the chance to grow into adult frogs to continue the life cycle.

Once our forests, meadows and hedgerows were full of beautiful wild flowers. Because they are so lovely to look at they are tempting to pick and, as a result, places once called Bluebell Woods no longer have bluebells growing in them. Admire the wonder and beauty of nature, but the only things that you should ever take are photographs!

Identifying animals and plants

If you are out walking, especially if you are visiting a wild life area, a wood or country park, take with you some pocket-sized spotters' guides to help you to identify the different plants and animals that you may see.

Carefully and closely looking at things is called observing. Observing important details about the things that you find will help you to classify them. That is, you begin to be able to group certain plants or animals because of things that they have in common. For example all trees that lose their leaves in the Autumn are called deciduous trees, those that don't are conifers or evergreens. You may find lots of different little creatures. Their shapes may be quite different but if they all have six legs and three definite parts to their body they are

probably all from the family of insects. There are many more different kinds of insects than of any other family of living creatures. From huge, beautiful butterflies and moths to the humble ant, from enormous and frightening stag beetles to wasps, bees and ladybirds. They are all very different, but they are all insects.

Being able to classify things into families helps when trying to identify them individually. Knowing if a toad is an amphibian or a reptile helps in searching in the correct reference book to discover its full name and other useful information.

Observing, classifying and identifying plants and animals are very important skills for any biologist (a scientist with a particular interest in living things).

What do plants need to grow?

Even if you do not have a garden it is possible to discover what plants need to grow strong and healthy.

You will need a pack of suitable seeds such as runner beans or french beans, sweet corn or cress. These begin to grow quickly so that you can follow their progress without having to wait for a long time. You will also need something to grow the plants in. You can use small flower pots, but empty yogurt cartons will do just as well, and a few jam jars and old, unwanted saucers will be useful. Also get some soil or a bag of seed compost and some cotton wool.

Begin by asking yourself what you think plants need to grow. Most plants that we see grow in the ground or in soil in some kind of pot. Do all plants need soil? How could you find out? Farmers often complain that the weather is too wet or too dry, that it is too cold or too hot. Is water important to a plant and does it matter how hot it is? Do plants prefer sunshine or darkness? There are a lot of questions here, so how should you begin?

Controlling your experiments

Scientists often use what they call a controlled experiment. For example, to see if plants prefer sun to shade some would be grown in darkness and others in bright sunlight. But to make the test fair all the other things would be carefully controlled. They would use the same seeds, same soil, same pots, same amount of water, same temperature, and so on.

Try this investigation for yourself.

Wet or dry

Get three saucers or small plates, some cotton wool and some mustard and cress seeds. Put about the same amount of cotton wool as a pad on each saucer. Sprinkle some cress seeds evenly over each pad and place them side by side on a warm, sunny window sill. Leave one saucer dry, make the second damp but the third should be covered with water.

Check each day to see how things are going. Keep the first saucer absolutely dry, add just enough to keep the second damp to the touch but the third should be swimming in water.

Very soon some of the seeds will begin to grow. Watch carefully to see how the tiny plants begin to develop. What grows first, the shoots or the roots? The shoots will be yellowish to begin with and then turn green. The roots are creamy white and grow very long with lots of fine root hairs sprouting from them.

Watch the progress of your plants over a number of days. How well do you think the seeds on the dry cotton wool will grow? What about the others, are some of the tiny seedlings growing better than the others? It is impossible to tell exactly what will happen to your plants but too much water can be as harmful as too little.

How water moves up a plant

Water is very important to plants. They cannot grow without some water. The roots, covered in tiny root hairs, take in water which passes up the plant through the stem to the leaves. You can actually see this happen in a very simple way.

Get a freshly cut flower such as a daffodil or a rose and place the stem in a jar containing water with some ink or food dye in it. If you can't get a flower try a fresh stick of celery instead. Leave the flower for a day then cut through the stem about half way along its length with a sharp knife or scissors. Look closely at the cut stem (if you have a magnifying glass you will see what has happened even more clearly). There are special cells in a plants stem

that carry the water up into the leaves and flowers. You can see them because the ink or dye stains the cells making a ring of colour in the stem. In the celery you will see lots of little coloured spots marking where the water has passed up the celery stalk.

It is even possible for the coloured water rising up the stem into the flower to change the colour of the flower as well. Look carefully at the tiny veins running through the petals, they will look darker because of the dye in the water.

Do fertilizers really work?

There are lots of adverts for fertilizers that claim to make plants grow bigger and stronger with more flowers or fruit. How could you find out if they work? If you tried the experiment with cress seeds you will have found that some of them grew very well but they all finally died off. This is because plants need more than just pure water to grow. Most plants are grown in some kind of soil. A plant gets all the food that it needs from the soil, so do fertilizers help?

Setting up the test

Get some seeds, such as French beans, runner beans or corn. You will also need lots of pots, empty yogurt pots will do, some potting compost or soil and one or two small bottles of liquid fertilizer.

Soak the beans or seeds overnight in fresh water to help them germinate or start growing. You will need about 15 seeds and enough pots and soil to grow each one separately. Treat each seed in the same way. Fill each pot three quarters full with compost or soil and push a seed into each pot. Sprinkle a little more compost over each to cover the seed. Place all

the pots in the same place so that they will be at the same temperature and get the same amount of sunlight.

Mix the fertilizers according to the directions on the bottle and put one of the mixtures on five of the seeds, the other mixture on another five seeds and plain water on the remaining five seeds. These seeds will be the control group. Use a measuring jug to make sure that each pot gets the same amount. Label each pot so that you remember what to put in each of them. Be sure to water them when the soil begins to get dry, but always use the same amount and use the correct mixtures for each group of pots, with only plain water for the control group.

Keeping growth records

As the plants begin to grow measure each one every 5 days and keep a careful record of their height. After a few weeks bean seeds may need some support as they begin to grow tall. Use thin cane or string for the plants to curl around. After several weeks measure each plant and add together the heights of each set of five plants and divide by 5 to get the average height of each group. You may need to use a calculator to help you. Is there any difference in the averages? Is any one group doing better than the others? Repeat this from time to time as they grow. Do the plants grown with fertilizer grow taller and stronger?

Whatever your results it is an interesting and important experiment and by trying it you will learn a great deal about how plants grow and what they need to keep them strong and healthy. Perhaps you can put all your new found knowledge to good use by growing a variety of plants in a small garden plot. You can buy a packet of mixed flower seeds in most garden shops and you will get so much pleasure from looking at the colours and shapes of the plants that you grow.

Natural decay

You have learned a great deal about how things grow, but what happens to them when they die? A large plant has lots of valuable food locked away inside it. If an animal eats the plant it is broken down inside the animal's body releasing the food which the animal can then use for energy and growth.

A plant that dies begins to rot or decay. The parts that make up the plant begin to break down and other plants and living things can use the decaying plant as food for themselves. In nature nothing is wasted, but is always used again or recycled as it is called.

Watching the rot set in

If you have a few glass jars, such as empty coffee jars, with screw top lids you can watch things that are rotting or decaying. Try some fruit that is past its best. We eat fruit as food to build healthy bodies and give us energy, but watch carefully what happens to the fruit when it is left for some time. Place an apple or pear, an orange or banana or any fruit that you may have into a jar and screw on the lid. You could also try other foods such as a few slices of bread.

It is important to keep the lids on for two reasons, firstly, whatever happens to the food must have been caused by things that are already inside or on the surface of the food and secondly, it is wise NOT to handle rotting food. This way you can watch it safely without getting anything on your hands.

You will need to leave the fruit or food for some weeks, but check it from time to time to see what changes are taking place. Many of the things will certainly change colour and perhaps shape too. You will find strange things growing on the surface of some of the food. Bread often gets covered in a blue mould and you may notice the same kind of thing on the skin of an orange. Others will grow whispy 'beards' of white mould.

Eventually, after a long time, the fruit will break down into a dark brown mushy lump. At this stage there will be little more of interest to see. If the jars are no longer needed it may be best to throw them away, otherwise throw the rotting fruit into a hole dug in the ground and cover it over. Wash out the jars very thoroughly, and remember to wash your own hands too.

Food chains

Many things will eventually rot away if left long enough, it is nature's way of returning precious raw materials to the ground to be used again. New plants grow on the remains of dead and decayed material. Some of the plants are eaten by animals which may get eaten themselves by other animals.

For example an oak tree, that grew from an acorn that fell onto the leaf mould of a forest floor many years before, sheds its leaves every Autumn. Tiny creatures, insects and worms live on the decaying leaves. A vole might catch and eat the worm only to be caught by an owl or a kestrel that feeds its own newly hatched chicks. Everything eventually dies and decays to begin the cycle all over again.

On a nature walk

Look for evidence of plants or animals feeding off others. In a forest find a fallen tree. If it has been down for some time you will find all sorts of things living off the dead wood. Look for tell-tale signs of wood-boring beetles, little holes and tunnels running through the wood. If the bark of the tree is loose, peel some away. On the back you may find a maze of tiny tunnels made by bark beetles. In dark, damp spots look

for wood lice feeding off the rotten wood. Other things to look for are the many different kinds of fungus that grow on dead wood. Large bracket fungus, like shelves, wrapped around the bark and perhaps toadstools as well.

You may spot grey squirrels scampering over the ground or leaping up the trunks of trees. Look for the remains of nuts, their husks and shells, where squirrels have been cracking and eating them. You may spot birds pulling worms from the ground after heavy rain, when the worms have come close to the surface, and look out for a thrush breaking the shell of a snail by dropping it and banging it against a stone.

Owl pellets

You may be lucky enough to spot an owl's perch in an old tree or a barn by finding an owl's pellets on the ground. When owls catch their prey they tend to swallow them whole, fur, bones and everything. The bird can digest most of its dinner but it will get rid of the things that it can't by bringing them back up again in the form of a furry pellet. If you find any you can pull them apart with tweezers and find the skulls and bones of tiny animals such as mice and voles inside.

Every living thing, even ourselves, is part of a food chain where each thing in the chain feeds on something else. All food chains begin with plants which are the ONLY living things that produce new food for the food chains, created by the action of sunlight on their leaves. This very important process is called photosynthesis (say 'photo - sin -thu - sis').

18

Solids, liquids and gases

What a load of rubbish

One of the major problems of human life is knowing what to do with all of our rubbish. Just look at how much gets thrown away each day by one household, your own. Every week you put out plastic sacks or a dustbin that is full of rubbish. This will probably include packets and tins, uneaten food scraps, newspapers and all kinds of wrappers, paper and plastic, glass bottles and even the occasional broken toy.

We are becoming more and more aware of how much is wasted. Every newspaper, wrapper or card package is made from paper that comes from wood pulp made by cutting down trees. It takes years for trees to grow, yet thousands and thousands are cut down every day just to make paper.

Plastics are very important. Many things could not be made as easily, or as cheaply, if it was not for plastics. But a great deal of plastic is made simply to wrap things in, such as bags or packets. Many plastics are made from oil which was made deep under the ground millions of years ago. Oil cannot be replaced so, when it runs out, so does the plastic.

Metals, such as aluminium drinks cans, need a great deal of energy to make. Yet all we do is throw them away.

Recycling our waste

What can be done? People are getting very concerned about unnecessary waste and the idea of recycling our rubbish is becoming more popular. You may have seen in some towns places called bottle banks, where glass bottles can be left to be remelted and moulded into new glassware. There are also a few can banks for aluminium drinks cans as well. Many schools and organisations save newspaper which can be pulped down to make more paper and card.

In some countries rubbish is burned and the heat from the furnaces is used to keep blocks of flats warm in the winter. But even if we try to recycle as much as we can, we will still be left with mountains of rubbish to get rid of, and this is where another problem arises.

Bio-degradable

One of the easiest ways to dispose of rubbish is to dig a large hole in the ground, fill it with rubbish and cover it over again. This would be fine if it eventually rotted away, returning the raw materials to the soil again, but many new man-made materials do not rot.

Collect some typical examples of rubbish such as newspaper, a can, food scraps, a plastic bag, polystyrene packing material, cardboard, a glass bottle and perhaps some things like a piece of wood, a few nails and a piece of broken crockery.

For this investigation you will need a lot of patience. Find a suitable spot in the garden, dig a shallow trench and bury your items of rubbish. Put some kind of label in the ground above the spot to remind yourself where the things have been buried. Make a note of the date too.

Leave the things undisturbed for as long as you can, several months at the very least, then carefully remove the top soil to see what has been left behind. What has happened to such things as the paper, or card? Does anything remain of the food scraps?

What about the glass, plastic or crockery. While many of the things will still be there, some of them will have changed. What about the wood or the iron nails, are they still as good as the day they were buried? Is the can still a bright silvery colour?

You will discover that some things rot away, other begin to rust but some of the things hardly change at all. Because of the problems of getting rid of rubbish, scientists are finding ways of making many more materials such as plastics that will rot. These new kinds of plastics are called bio-degradable which means that they will biologically change and decay away.

Testing materials

Many scientists spend a lot of time making and testing different materials. These might be different kinds of plastic or metals, paint, glass or even soap powders. They have to find out if the thing that they are testing will do its job well. After all, a new plastic for lemonade bottles would be no good if it cracked easily and a new soap powder would not sell very well if it did not get rid of stains.

Here are some tests for you to try.

Which is the best carrier bag?

For this test you will need a strong wooden pole such as a broom handle, sand or soil and a small spade, bathroom scales and as many different carrier bags as you can find (like the ones given away free at supermarkets).

Push the pole through the handles of one of the carrier bags and lay the pole between two stools or chairs so that the bag is about 15cm (6 inches) above the ground. It is a good idea to cover the floor under the bag with layers of newspaper to save making a lot of mess!

Begin to fill the bag with sand or soil. Watch the handles and the bottom of the bag carefully for signs of stretching or tearing. You will be surprised how much sand a bag can take, probably more than you could carry, so keep your toes well away from the bag in case it suddenly snaps and falls. Add the sand slowly and carefully, particularly if you feel that the bag is on the point of breaking. When it finally gives way either the handles will snap or the bottom will burst.

Measuring the load

Put all the sand or soil that was in the bag, including any that may have spilled when the bag finally broke, into a large bowl or a bucket that is resting on some bathroom scales. Make sure that you have set scales to the zero mark before putting in the sand. If you are not sure how to do this ask an adult to help.

Write down on your test result sheet the weight of the sand in kilogrammes and then do your test all over again with a new bag. After two or three goes you will be getting good at guessing or estimating how much sand a bag might take and at the end of your tests you will be able to recommend which bags are the stongest and which ones not to use.

An egg tester

Take an ordinary chicken's egg and hold the points between your finger and thumb. Now squeeze it as hard as you can. You may wish to do this over a sink or bowl but, so long as there are no cracks or faults in the shell, you will not be able to break it because egg shells are very strong.

To find out just how strong they are try this test. You will need an empty washing up liquid bottle or a small plastic lemonade bottle, sharp scissors, some plasticine or modelling clay, a small block of wood and lots of heavy things to act as weights.

Cut the top off the bottle about a quarter of the way down, place a small cushion of plasticine in the end of the funnel shape and wedge this inside the rest of the bottle. Sit the egg on the plasticine so that it is standing perfectly upright and rest the wood block on the

Cut here with scissors

Egg sitting on plasticine cushion

Block of wood balanced on top of egg using a cushion of plasticine

Funnel shape pushed upside down into bottom of bottle

Adding large metal weights to the pile on top of the block

top of the egg. Use another small cushion of plasticine so that the wood rests squarely and firmly on the egg's point. It is a good idea to choose a wood block that just fits inside the bottle so that it will not be able to wobble too much.

Now start placing heavy, flat weights on top of the wood block. Rest them on gently, don't drop them. You may find that you run out of weights long before the egg breaks. Try balancing house bricks on the pile, but remember to keep your feet out of the way in case they fall off. So long as the egg remains upright and is taking the weight on its points it will support an enormous load. If your egg finally gives way, place all of the weights on some bathroom scales to see how much it took.

It is possible to drop an egg on to a springy patch of long grass from quite a height without breaking it. From what height does your egg fall and still survive? Try it and see.

Things that dissolve

Have you ever watched someone putting sugar in a cup of tea? They take a spoonful of solid, white crystals of sugar and stir them into the hot tea. When they drink the tea there are no lumps floating around, so where has the sugar gone? Take a glass and pour some warm (NOT HOT) water into it. Taste the water. Now stir a spoonful of sugar into the water. Watch the sugar carefully, it seems to disappear, but taste the water again and you can taste the sweetness of the sugar. It is still there, but you can't see it, we say that it has dissolved. The liquid that is left when something has dissolved in water is called a solution.

Finding the solution

What other things dissolve in water? Find as many different things as you can to try. Choose things like flour, coffee granules, gravy powder, salt, custard powder, soap flakes, crushed cereal or tea leaves. BE CAREFUL, don't try things that may be harmful, always check with an adult first.

Start with a small amount of cold water and add a level teaspoonful of one of the powders. Then give it a stir. Make a note of the things that dissolve and those that do not. Now try again, but this time use water that is hot (NOT BOILING), stir well and look again. How many of the things that did not dissolve in cold water will

in hot water? Tea leaves colour the water, so something has dissolved, but the leaves are left behind so tea only partly dissolves. What about custard and gravy powder? What happens to the water when it cools?

Does sugar dissolve in cold water? If so how much? Add a level spoonful at a time to a glass of water and stir until it has completely disappeared. How many spoonfuls does it take before some is left undissolved in the bottom of the glass? Now take exactly the same amount of HOT water in a glass and try again. Does it take more or less sugar than the cold water?

Filtering

We can use a tea strainer to catch all the leaves and let just the liquid through. A tea bag does the same thing, keeping the leaves trapped inside while letting the hot water take up the colour and the flavour of the tea.

Scientists filter liquids to trap any tiny pieces of solids that might be floating about. They use special paper called filter paper which they fold into a cone shape and place in a funnel.

You may be lucky enough to be able to buy some filter papers from a local, friendly Chemist, but if not thin blotting paper will do. Cut the paper into a circle about 15 to 20cm (6 to 8 inches) across. Fold it in half, then in quarters. Open up the cone shape and place it in a plastic funnel. (The kind that you can get from a shop that sells wine making equipment). Another way is to use the special coffee filters made for coffee machines. They are not very dear and they come already cone shaped.

If you tried the investigation to find things that dissolved in water you will have found that some things only partly dissolve. You could pour

these through your funnel and catch the liquid in a bowl underneath, while any solids will be trapped by the filter paper. If you leave the liquid in a shallow bowl or dish, the water will gradually evaporate away and will leave behind the solid that had dissolved in the water. You will find that some things (such as salt and sugar) are left as crystals.

Separating sugar from sand

Try this trick on your friends.

Mix together a tablespoonful of sugar and a tablespoonful of clean sand. Ask your friends if they can separate them again. It is almost impossible by hand but there is an easier way. Just put the mixture into a bowl containing a little hot water and stir. The sand will not dissolve but the sugar will. Pour the mixture into a filter paper and the hot water containing the dissolved sugar will drip through, leaving the sand behind in the filter paper. If you want to get the sugar back, leave the sugary water in a shallow dish or saucer on a warm shelf or window sill. The water will slowly evaporate away in just the same way as puddles disappear or washing dries on a clothes line. The sugar is left behind in the dish.

Which washes whitest?

Many adverts for washing powders and liquids say how well they work, even at low temperatures, but how good are they really? This investigation allows you to compare various products for yourself. You will need some pieces cut from an old, unwanted cotton sheet about 30cm (1 foot) square or a few old handkerchiefs, some bowls large enough to take 2 litres (about 3 pints) of water and some samples of different washing powders and liquids.

To make your test fair you must try to do everything in the same way. Prepare your pieces of material at least a day in advance. Stain each piece using a variety of things. For instance you may wish to rub one corner of each piece of material on a patch of grass, put a blob of tomato ketchup in another corner, blackcurrant fruit drink in another and rub melted chocolate into the fourth corner with, perhaps a squirt of ink in the middle. Leave the material to dry overnight, the material will now be very badly stained.

Pour the same amount of powder or liquid into each bowl (about an eggcupful should do) and top up with exactly the same amount of water at the same temperature (hand hot). Mix each powder or liquid to make sure that it has all dissolved then place a separate piece of stained material into each of the bowls and push them well under to be sure that each one is thoroughly soaked.

Leave the bowls for an hour stirring each one 5 times clockwise then 5 times anticlockwise every five minutes. At the end of the hour rinse each piece under clean running water and hang them out to dry. Remember to make a note of which material was washed by which powder. When they are dry compare each piece. Which was the hardest stain to get out? Are there any stains that all the detergents managed to shift? Did any of the detergents remove all the stains? Which, in your opinion, was the best?

25

The strength of shapes

One of the secrets of success when designing something like a new aircraft or a bridge, a car or a tall building, is to use the least amount of materials as possible to keep down weight and cost but still to make the thing as strong as possible. This is done by putting the parts together using strong shapes.

What is a strong shape? This investigation will help you to find out. Cut some strips of stiff card from the side of a strong cardboard box. Make them about 20cm long by 3cm wide (8 inches by 1 inch). Take 4 strips and join them together to make a square by pushing a paper fastener through each end. Take another 3 strips and join them together in the same way to make a triangle.

Testing their strength

Stand the square on its edge and gently push against one of the corners. You will find that the square is very easily pushed out of shape and is soon flattened. Now stand the triangle on its edge and push down on the point. The triangle is a rigid shape and can take far more pressure before it loses its shape. A square or rectangle can be made rigid by putting in a crosspiece between two diagonal corners. Try it yourself. Cut another strip of card 30cm (12 inches) long and, using the paper clips, fix it in place. Now when you press against a corner the square keeps its shape.

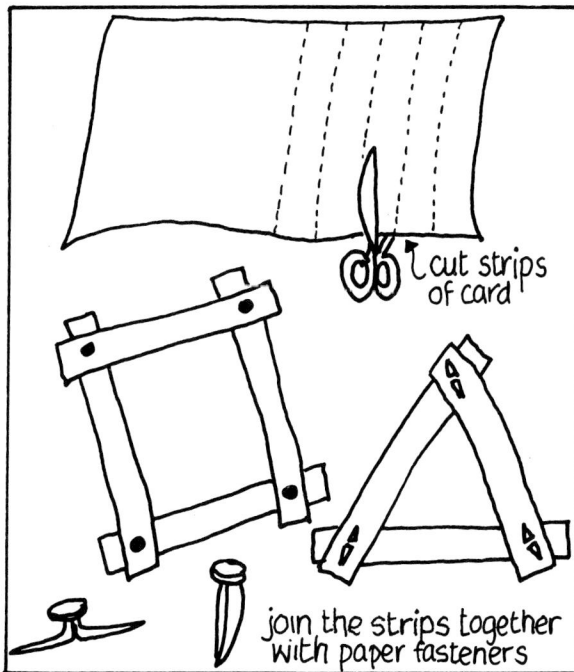

cut strips of card

join the strips together with paper fasteners

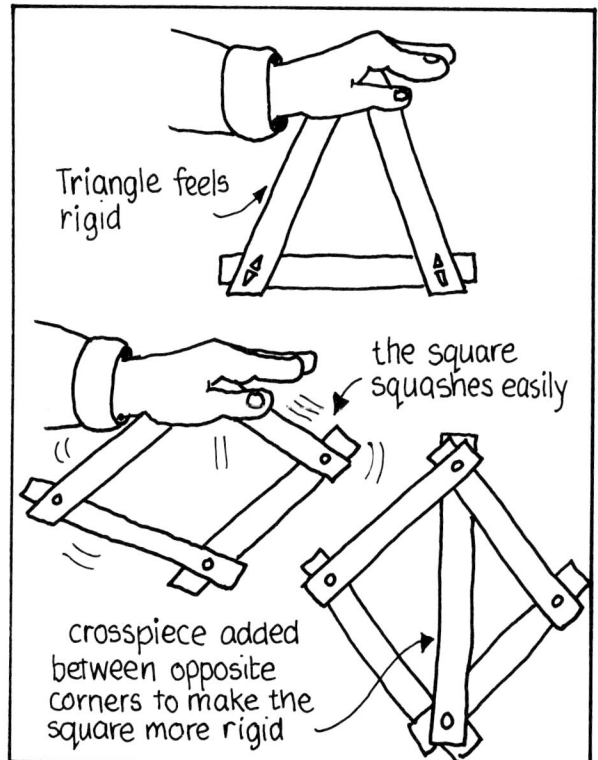

Triangle feels rigid

the square squashes easily

crosspiece added between opposite corners to make the square more rigid

Using strong shapes

Look at an electricity pylon. It is made from thin metal strips but, because it is bolted together to make lots of triangle shapes it is very strong.

Build your own pylon. To test yourself to see if you understand about strong shapes, try this challenge. Take 30 plastic drinking straws and a packet of pipe cleaners. Cut the pipe cleaners into suitable pieces to make joints to hold the straws together. You can twist them together to make 2, 3 and 4 way joints. Your problem is to make the tallest tower possible. A tower that is one metre tall is very good, two metres is brilliant, but it must stand up by itself, so think carefully about your design.

Pipe cleaner joints to hold straws together

If you find that this challenge is quite simple, here is another. Still using only thirty straws and pipe cleaner joints, build a tower that is strong enough to support the weight of something about the size and weight of a small toy car. It is not as easy as it sounds.

Bridge building

Triangles are not the only strong shapes. Ancient bridge builders found other ways of supporting huge weights safely. For this experiment you will need two house bricks or two piles of heavy books. Cut four strips of thin card from the sides of an old cereal packet. Make them about 30cm long and 10cm wide (12 inches by 4 inches). Lay two of the pieces, one on top of the other, across the gap between the books or bricks. Make the gap about 25cm (10 inches). Gently place a number of toy cars, stones or any other suitable weights onto the card bridge. How many of your weights does it take before the bridge collapses?

Now take the other two pieces of card and lay them across the gap, only this time wedge the ends of the bottom piece of card against the inside of the books or bricks to make an arch. Lay the second strip over the arch.

Now start loading this bridge with your weights. Even though you have used the same amount of card, because of the arch shape, the second bridge is much stronger.

Paper bridges

Can you bridge a gap with just a single sheet of paper? Lay a sheet of paper across two books and place some small weights such as coins on the paper. How many can it support before the bridge collapses?

A flat sheet has hardly any strength at all but if you fold it or roll it the sheet takes on a stronger shape and can support far more weight. Try folding the sheet into a fan shape or roll it into a tube. Use coins as weights to see which one of your paper bridges is the strongest this time.

Look closely at a bicycle, it is a perfect example of strength without unnecessary weight. If a bicycle was made from solid metal bars it would be incredibly heavy. Thin steel tubes are used instead to make the bicycle as light as possible. The tubes are still very strong and the bicycle frame is made stronger still by putting the tubes together in triangle patterns to make them rigid.

How do we know air is there?

We are told that air is all around us but we can't see it or taste it or smell it, how do we know that it is there? Usually we are not aware of air. It is not like wading through water which can be very hard work because water is very heavy. Does air weigh anything at all?

You can prove that air has weight with a very simple experiment. You will need a long, thin garden cane, two balloons and some cotton and Sellotape.

Blow up the balloons so that they are roughly the same size and shape and tie them to the ends of the cane with some cotton. Hang the cane by a piece of cotton tied to the middle.

If the cane does not balance straight and level, move the centre cotton away from the lower, heavier end and towards the lighter, higher end until it exactly balances. Now all the weight on the left side of the cane is exactly balanced by the weight on the right. Tape the cotton in place so that it cannot move.

Upsetting the balance

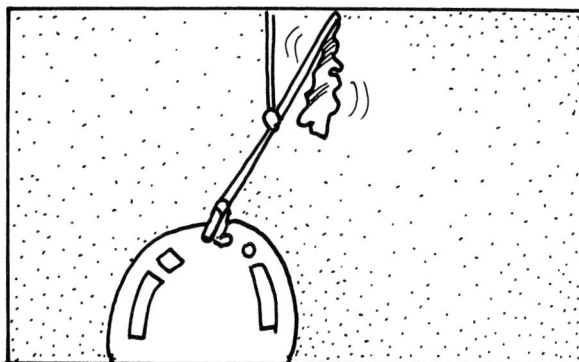

Burst one of the balloons with a pin, but make sure that none of the balloon's skin is lost. Look at the balance now. The only difference is that the air in one of the balloons has been lost but, because the other balloon is still full of air, it is much heavier and drops downwards.

We are only aware of air's weight when air moves. Wind is moving air. Sometimes air can move very quickly, such as in a tornado or a hurricane, then the force of the moving air can be very destructive, flipping cars over on their roofs and blowing down houses. When you pedal a bicycle fast down hill you can feel the force of the air against your face trying to hold you back. That is why so much research goes into streamlining cars and aircraft to allow them to slip through the air more easily.

The life giving gas

When we breathe we take in air. If we could not breathe we would suffocate and die. Fire needs air to burn. Smother a fire and stop air from getting to it and it will go out. But although air is a mixture of many different gases. it is only one of these gases, that keeps us alive and keeps fires burning. It is called oxygen.

Ask an adult to help you with this experiment. Stand a candle or night light on an old saucer. Fill the saucer with water. Ask an adult to light the candle for you and, when it is burning well, cover it with a jam jar. Watch what happens carefully.

The candle continues to burn but soon begins to splutter and then goes out. At the same time some of the water in the saucer is sucked up inside the jam jar, why is this? As the candle burns it is using up the oxygen in the jar. No more air can get into the jar so when all the oxygen has been used up the candle goes out. The water is sucked into the jar to take the place of the gas that has been used. You will notice that only a small amount of space is filled with water. This is because only a small part of the mixture of gases that make air is oxygen (about one fifth).

Where does oxygen come from?

If all living creatures are continuously using up the oxygen in the air and fires are burning it up, how does it get replaced? We rely on the green plants on land and the green algae in the sea to do this job for us. Plants make their own food by a special process called photosynthesis. They use sunlight and a gas called carbon dioxide to make simple sugars and, as a result of the chemical reaction, the plants give off oxygen.

You can watch this happening for yourself. Get a large bowl full of fresh water, a funnel, a handful of pond weed and a small plastic bottle full of water.

Put the weed into the water and cover it with the funnel, make sure that all the funnel is covered with water. Turn the full bottle upside down in the bowl and put the mouth of the bottle over the spout of the funnel.

Leave the bowl in bright sunlight for a few days. You will see little bubbles of gas forming on the weed which bubble up inside the bottle. The gas is oxygen being given off by the pond weed. As it collects in the bottle it forces the water out until you have several centimetres of gas in the bottle.

Splitting water

Water is one of the most common substances on Earth. Our own body is made up of 90% water. No living thing could exist without it. We take water for granted, but what is water made from? Get a powerful 9 volt battery, two wires, a deep bowl full of water and two small plastic bottles.

Join a wire onto each of the battery terminals and put the other ends into the water in the bowl. MAKE SURE THE ENDS DO NOT TOUCH. Pure water does not let electricity pass through it very easily but you can get it started by adding a little acid in the form of half a cupful of vinegar. This allows the electricity to pass between the wires and gas begins to bubble from the ends.

This is called electrolysis. The electricity 'splits' the water into the two substances that make up water; these are the two gases hydrogen and oxygen. All things are made up of tiny building blocks called atoms. Different atoms can join together to form molecules. A molecule of water is made of two atoms of hydrogen with one atom of oxygen.

Fill the bottles with water and turn them upside down in the bowl. Push the end of a wire into the neck of each bottle. Within a short time the gases bubbling up into the bottles will begin to force the water out. You will also notice that one of the bottles seems to have twice as much gas inside it (the hydrogen) as the other one (the oxygen).

Make a volcano

This clever trick looks like a volcano erupting, but it is perfectly safe to do.

Make a good sized volcano shape with a large ball of modelling clay or plasticine. Make your volcano hollow inside and fill it with baking powder (bicarbonate of soda) and a few drops of red food colouring.

Volcano-shaped plasticine hollowed out

RED

Pour in vinegar and the red baking powder froths and bubbles out of the volcano

To set off your volcano, put it on a tray and pour an eggcupful of strong vinegar into the top. The acid in the vinegar reacts with the baking soda making lots of carbon dioxide gas which makes the mixture bubble and froth out of the volcano. The red food dye makes the frothy mixture look like fiery lava pouring down the sides.

Testing for acids

Chemists use special liquids called indicators or paper that has been soaked in an indicator to find out how strong an acid is. One kind is called Universal Indicator which turns different colours for different acid strengths. Another well-known indicator is called litmus paper which turns from a blue colour to red even if only a few drops of weak acid fall on it.

You may not be able to get any of these indicators, but you can make your own. All you need is a small amount of red cabbage.

Ask an adult to help you to shred the cabbage and boil it in a pint of water in a saucepan for a few minutes. When it is cool, strain off the liquid into a jug or bottle. You will notice that the water has been turned blue by the cabbage, not red!

Pour a little of your blue cabbage water into a glass tumbler and add just a few drops of vinegar. Stir the water and you will find that it turns a clear red colour. The cabbage water acts like an indicator. Acids turn the blue colour red.

Now add a teaspoonful of baking powder to the glass. When you stir you will see some bubbles of gas form. This is because it is reacting with the vinegar making carbon dioxide. But you will also notice that the water goes back to a blue colour again because the acid has been used up (we say it has been neutralized) by the bicarbonate of soda. Things that react with acids in this way are called alkalis (say alk - a - lies).

Test other liquids to see if they are acids. Try fruit juices like lemon and orange, lemonade, cold tea and milk. Some things may have acid in them and turn the blue water red, others will not.

Wind, rain and the weather

How much water in a puddle?

If it rains look for a large puddle on a flat area of concrete or tarmaced surface. When the rain stops take a stick of chalk and draw round the edge of the puddle. Leave the puddle for exactly 15 minutes and draw round it again. Keep doing this until all the water has evaporated away. Get a measuring jug and some water and refill the dip to the first chalk line. Keep a careful note of how much water it takes to do this, measure the amount in litres.

Using the jug, refill the puddle to each of your chalk lines, keeping a record of how much water it takes. If you add together all the amounts you will know how much water there was in the puddle and how long it took to evaporate away, but what is more interesting is to see when the water was evaporating fastest. In which period of 15 minutes did the largest amount of water evaporate away and which was the smallest. Does a puddle evaporate faster when it is sunny or windy?

Measuring rainfall

You may have heard on weather forecasts that an inch of rain fell in two hours during periods of very heavy rainfall. But how much rain falls in a week and which is the wettest month? By making your own rain gauge and keeping careful records you should be able to answer these questions and more.

To make your gauge you will need an empty plastic lemonade bottle, a ruler and a pair of scissors.

Using the scissors, cut the top off the bottle to make a funnel which you wedge into the neck of the remaining part of the bottle.

Place your rain gauge in a suitable spot in the garden, you may need to surround it with bricks or half bury it in the ground to stop it from being accidentally knocked or blown over. Make sure that it is not sheltered by a wall or under the branches of a tree as these could give you false readings.

At the same time each day remove the funnel and dip your ruler inside. If it has rained some water will have collected in the bottle. Pull out the ruler and see how far up it has been wetted. Record the amount to the nearest millimetre and reassemble your gauge ready for another day. You can add up the total rainfall in a month and divide by the number of days in the month to get an average daily rainfall. If necessary, you can use a calculator to help you to do this.

Cut here

Top wedged into lower part of the bottle

Make your own weather vane

A weather vane is useful to see which direction the wind is blowing from. Our weather conditions often depend upon wind direction. A North wind coming from the icy northern areas can bring cold, wintry weather.

To make your wind vane you will need an old wire coat hanger, a pair of pliers, a drinking straw, the jet nozzle from the top of a washing up liquid bottle, a hammer and some wire staples, a piece of stiff card cut from a washing powder packet, scissors and Sellotape.

Cut out a large arrow shape from the card about 40cm (15 inches) long.

Tape the straw to the side of the arrow just in front of the half way point along its length. Use the pliers to cut a piece of wire from the coat hanger about 30cm (12 inches) long and

staple it to a suitable fence post or a wooden stake knocked into the ground. Place the jet nozzle over the wire and then put the arrow in place. The plastic jet allows the arrow to swing freely in the wind.

The only other thing you may need to know is the direction of North so that you can mark the compass points on your post to be able to read the wind direction.

If you do not have a compass turn to page 49 and find out how to make your own.

REMEMBER when you record the direction of the wind your arrow will be pointing in the direction that the wind is COMING FROM. So a West wind is coming from the West, a North wind is from the North and so on.

Wind speed indicators

It is important to have some way of recording the wind speed as well. The speed of the wind can affect your body temperature enormously. On a still, sunny day a temperature of just 15°C. (about 60°F.) can feel pleasantly warm. But if the wind is blowing quite hard it will feel very cold. This is called the wind chill factor.

Here are two ways in which you can measure and record wind speed.

The first is very simple. All that you need is a piece of card about 30cm long and 15cm wide (12 inches by 6 inches) cut from the side of a cereal packet. Sellotape a plastic drinking straw to the top edge. Find an unwanted wooden box, such as an orange crate, at a market or nail a piece of wood to a wooden base. Use a pair of pliers to unbend a wire coat hanger and cut a piece about 20cm (8 inches) long. Use staples and a hammer to fix the wire to the side of the

box or the top of the upright piece of wood so that it sticks out sideways. Slip the drinking straw over the wire so that the flap of card swings freely.

To use your wind speed indicator, place it in a suitable unsheltered spot facing any oncoming wind. The wind will blow the flap up at an angle. The stronger the wind the greater the angle. Attach a piece of card to the wooden frame or box using drawing pins. Mark the card at various angles with a ruler and pen.

A light breeze may blow the card to an angle of only 10 degrees, while a strong wind may blow the flap to 30 or 40 degrees from its normal, vertical position. After a few weeks you will get good at estimating from the angle of the flap how hard the wind is blowing. When you keep your weather records you will know that an angle of, for example, 70 degrees means a strong gale but only 20 degrees is a light wind and so on.

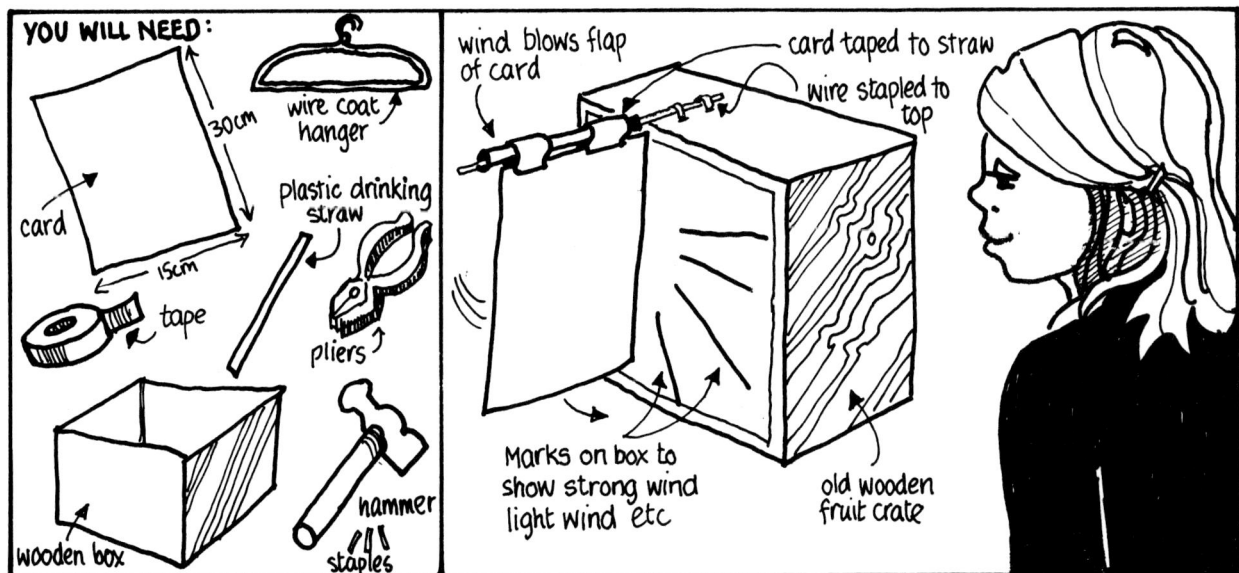

A home-made anemometer

YOU WILL NEED:

Four plastic cups or yogurt pots

a bamboo stick

Two long thin garden canes

a nail

a washing up liquid bottle

cut here

nail through jet nozzle into the bamboo stick

two garden canes pushed through holes in funnel shaped top

colour one cup to make counting easier

You may have seen an anemometer. This is an instrument used by weather stations to record wind speed. It looks like three or four little cups, the size and shape of half a tennis ball, on the end of short rods. The rods stick out from a central spindle so that, when the wind blows, the cup shapes are blown round. The speed of the wind is worked out from the number of times the anemometer spins round in a minute. This is often done fully automatically by connecting the anemometer to a special gauge that shows the wind speed in miles or kilometres per hour.

Obviously a professional instrument of this kind is very expensive, but you can make your own for nothing.

Collect four plastic cups or yogurt pots. Two long, thin garden canes, a bamboo stick about one metre long (or any similar long, thin stake of wood) a nail about 5cm (2 inches) long and the funnel shaped end cut from the top of an empty washing up liquid bottle.

Knock the bamboo cane or wooden stake into the ground. Make holes in the plastic funnel so that you can push the canes through to make a cross. Then make small holes in the cups and push them onto the ends of the sticks.

Push the nail through the small hole in the top of the funnel. Make sure that the nail is thin enough for your anemometer to spin freely. Tap the nail into the top of the cane or stake. To finish your anemometer, paint one of the cups with any available colour, this helps to make counting revolutions much easier.

With your anemometer set up all you have to do is count the number of times that the top turns in one minute. A simple way to do this is to count the number of times the coloured cup

passes you while someone times fifteen seconds with a watch. Multiply the number by four to calculate the number of revolutions in a minute. Keep a note of these totals with your other weather records. You will soon get to know the number of turns that represent a light breeze and the number of turns that you can expect in a strong wind.

Temperature and pressure

The other information that is important for weather records is temperature and pressure. High air pressure can be the sign of good weather whereas low pressure can bring rain and storms. Some people have air pressure instruments as ornaments, they are called barometers. If you or a neighbour has one, use the information that it gives for your weather records.

Temperature can be read from a thermometer. If you have a thermometer, place it in a shaded spot (not in bright sunshine) and take the temperature at the same time each day, or perhaps twice a day, morning and evening. If you are very keen on setting up your own weather station you may want to get a special kind of thermometer called a maximum - minimum thermometer. This is set up permanently in a good spot in the garden and it records the highest temperature of the day and the lowest temperature too. Little pointers move up and down the temperature scales and mark the temperatures reached. When you have recorded them, you reset the markers ready for the next twenty four hours.

Keeping weather records

To make the best use of all your weather instruments it is essential to keep a full daily record of all the information in a log book. Make a note of rainfall, cloud type, the amount of cloud cover, pressure and temperature (if you can), wind speed and wind direction.

If you do this for a good period of time you will begin to notice patterns in the weather. You may even be able to predict what kind of weather to expect. Compare your weather records with local daily forecasts on radio or television to see how close your readings are to those made by the professionals.

Wind power

Moving air has been used to power things, by pushing against them or turning them round, for thousands of years. Sailing ships that are blown by the wind can be little yachts for one person or huge cargo ships full of goods weighing hundreds of tons.

Moving air can be very powerful, and very destructive. Ships caught in storms have to be careful that they do not put up too much sail or the masts might break, the sails may be ripped or the ship may even blow right over, which is called capsizing. On land, storms and hurricanes can tear roofs from houses and uproot large trees.

Windmills

Although most of the windmills in this country were made in the eighteenth and nineteenth centuries (the 1700s and 1800s) many can still be seen today, and some have been carefully restored to full working order.

The large blades, called sweeps, were often covered in canvas sails and, when the mill was turned into the wind, the power of the wind would fill the sails and turn the mill sweeps. These turned a long wooden shaft inside the mill which was connected by large cogs to heavy mill stones that turned round and round to grind the corn. Some large mills could turn as many as three sets of stones.

Make your own windmill

Although it would be difficult to make a proper working windmill, you can make something that will turn in exactly the same way, just using wind power.

Take a square of stiff paper about 20cm (8 inches) long. Fold it diagonally in half from corner to corner so that the folds from the corners cross at the centre. Use a pair of scissors to cut from each corner, along the fold lines towards the centre for just a little less than half the length of each line.

Ask an adult to help you to unbend a wire

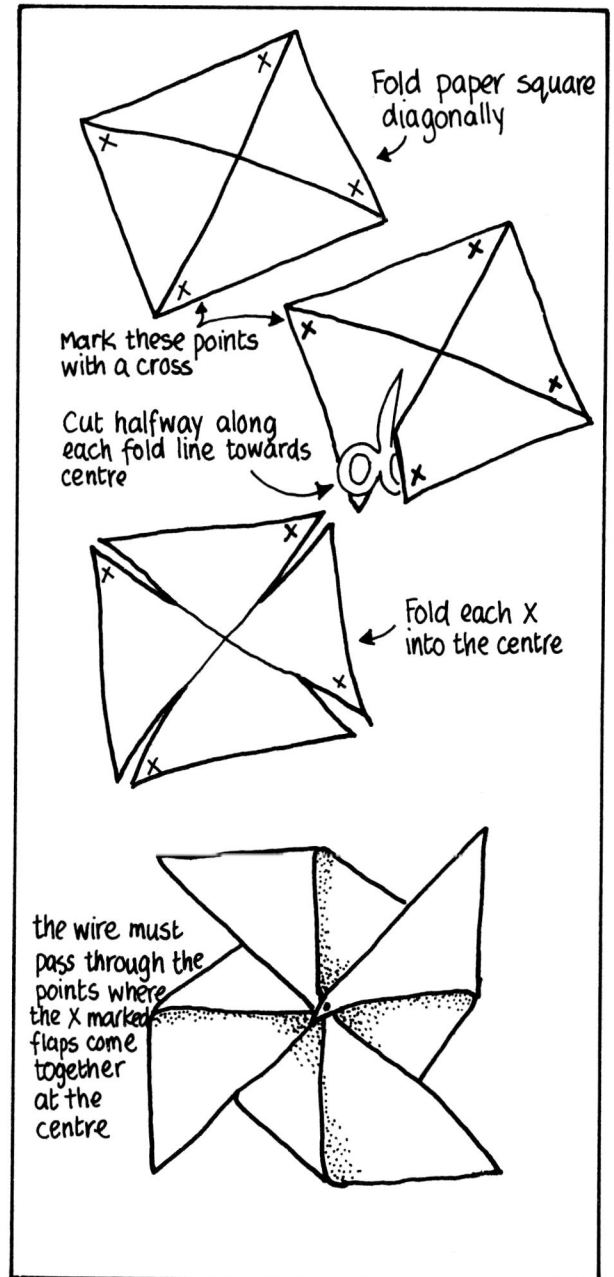

Fold paper square diagonally

Mark these points with a cross

Cut halfway along each fold line towards centre

Fold each X into the centre

the wire must pass through the points where the X marked flaps come together at the centre

40

Bent coat hanger and put nozzle

end of wire bent over

Washer or bead in place.

coat hanger and cut off a piece about 40cm (16 inches) long. Bend the top 10 cm (4 inches) over at right angles using a pair of pliers. Push the jet nozzle from an empty washing up liquid bottle onto the wire then push the wire through the centre of your paper square.

Take each corner, marked with a cross in the picture, and fold it in towards the centre. Make a small hole, roughly where the cross would be, using a needle, and push it onto the end of the

wire so that the four points overlap. Finally push a small play bead or small metal washer onto the end of the wire and bend up the end half centimetre of wire to stop the windmill from falling off again.

The jet nozzle at the back and the bead or washer at the front stop the paper windmill from catching on the wire handle and let it spin freely. Turn the windmill into the wind and it will spin very fast, even in a light breeze.

Land yachts and sailing ships

Wind can be made to blow things along. We usually think of sailing ships but some people also build metal frames on wheels with large sails that are blown by the wind along long sandy beaches. In cold countries some land yachts are made with skids instead of wheels, like large ice skates. These ice yachts can reach terrifying speeds of well over a hundred miles an hour, blown by strong winds across frozen rivers and lakes.

A simple yacht

Ask an adult to help you slice a medium sized potato into scallops about a centimetre thick. Use a cocktail stick as a mast and push a piece of stiff paper or thin card about half the size of a playing card over the stick as a sail.

Paper sail
cocktail stick
potato slice

Place your little potato boats in a large bowl, a bath or even a large puddle and blow them along. You can have races with a friend. Experiment with different sized sails to see if larger sails make them go faster, but be careful that you don't make them so large that they blow over. This is a problem that all yacht designers face. To use a large sail area the boat must have weight low in the water to balance it, this is called the keel.

Making a keel boat

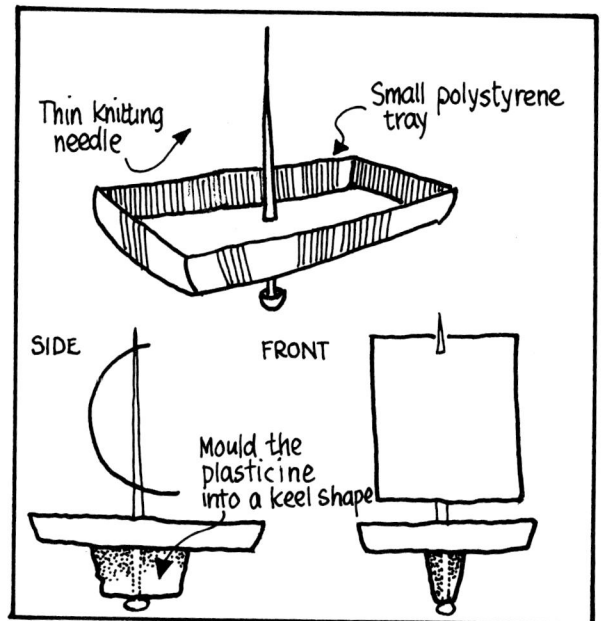

Thin knitting needle

Small polystyrene tray

SIDE FRONT

Mould the plasticine into a keel shape

Use a polystyrene tray (the kind that often contains meat sold in supermarkets) or a chunk of polystyrene packing material and push a knitting needle through the centre. Put a blob of plasticine or modelling clay on the end of the needle sticking through the bottom and shape it

42

to form the weighted keel of your boat. Now try different sizes of sail using paper or pieces of fabric to discover the best combination of sail size and keel weight to make the fastest yacht that is still stable in a strong breeze.

Thunder and lightning

Low air pressure and high winds can lead to the onset of stormy weather. One of the most spectacular weather conditions that we sometimes see is an electric storm or thunder and lightning.

Air, moving in a storm cloud, makes the tiny drops of water and ice particles rub together so that they become charged with static electricity.

If you are not sure how static electricity can be made try rubbing a balloon against a woollen jumper. The balloon becomes so strongly charged with electricity that it can be made to 'stick' to walls or the ceiling, held only by the static electrical charge. As an alternative hold the charged balloon over your head while you look in a mirror. The static charge attracts your hair making it stand on end.

A charge of static electricity builds up in a storm cloud until it is released by flashing from one cloud to another or from the cloud to the ground. We see the electricity jumping to the ground as a bright flash that lights up the sky. The crash of thunder that follows it is the noise of the lightning suddenly heating the air with the force of a violent explosion.

Electricity and magnetism

Electricity

Electricity is one of the most important forms of power that we use today. Without electricity we could not carry on living as we do. Almost everything we use either needs electricity to make it work or needs electricity for it to be made. The amazing thing is that when our Great Grandparents were children electricity was very new, houses and streets were still lit with gas lamps and cooking was done on open fires or gas stoves. Only a very few homes and offices had electric light and there were no microwaves, fridges, televisions, dishwashers or automatic washing machines that so many of us now take for granted.

You can learn a great deal about electricity with just a few simple items that you will be able to buy in any good electrical shop. You will need two or three small torch bulbs. Ask for ones that take between 3 to 6 volts of electricity to make them light up. Ask for some cheap bulb holders too, and a small screwdriver to undo the screws on the holders. You will also need to get some batteries. Get two small 1.5 volt batteries (AA size) that are often used to power toys and models, as well as a 4.5 volt battery used in bicycle lamps. Lastly, you will need some lengths of wire with the plastic insulation stripped off to leave about 2 cm of bare wire at each end. (You may need to ask an adult to help you with this.)

REMEMBER, ALTHOUGH IT IS QUITE SAFE TO EXPERIMENT WITH BATTERIES AND BULBS, NEVER PLAY WITH THE MAINS ELECTRICITY, IT COULD KILL YOU.

A simple circuit

Begin by screwing a bulb into a holder, take two wires and a single 1.5 volt battery. Can you find where to put the wires to make a pathway for the electricity to flow around so that the bulb lights up? Don't give up, try different ways, you will soon succeed. Always remember that

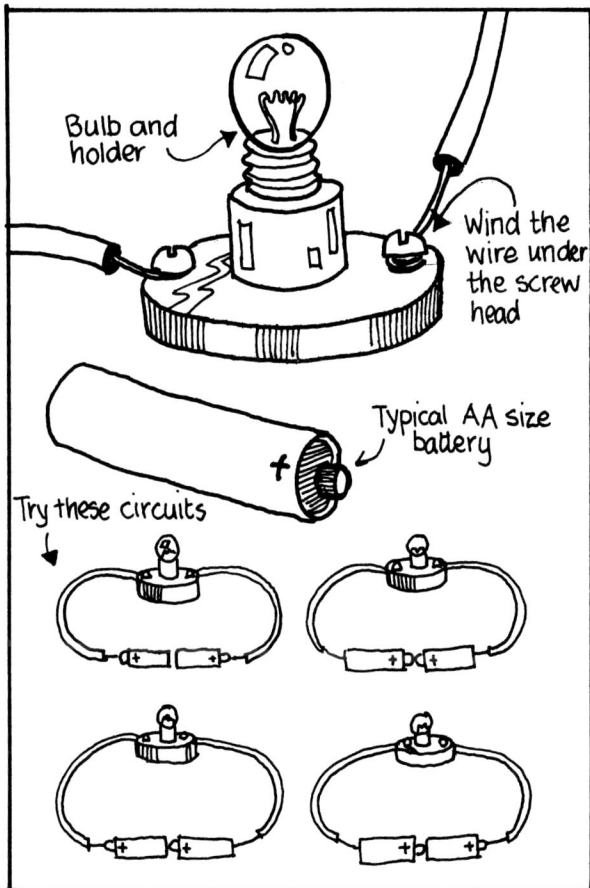

Making two bulbs light

To try this challenge you will need two bulbs in holders, some wires and a 4.5 volt battery. Join the wires to the bulb holders and try different ways of joining them to the battery to make both bulbs light up at the same time. There are two quite different ways to do this. When you have both bulbs alight look at how bright they are. If the bulbs appear quite dim and, when you unscrew one of the bulbs, they both go out, you have made a **series** circuit. If the bulbs look bright and, when you unscrew one of the bulbs, the other bulb stays alight, you have made a **parallel** circuit.

electricity needs to flow from one end of a battery through the bulb and back to the other end of the battery to make a complete ring or circuit.

When the bulb lights up you may be disappointed that it only shines quite dimly. Now use two 1.5 volt batteries. Can you think of a way of joining them together to double the power? Try making circuits like those shown in the picture. Which ones make the bulb light and which do not?

Conductors and insulators

Wires make a pathway for electricity to flow from a battery to a bulb and back again. We say that the wire conducts electricity. Electricity can also flow through many other things. Here is a simple way to discover some of them. Set up a circuit like the one shown in the picture.

You will find that the bulb does not light up because the electricity cannot jump the gap between the two drawing pins. Find as many different things as possible to lay across the two pins to see which ones light the bulb by letting the electricity flow through them and around the circuit. Try such things as a nail, your screwdriver, a glass, a plastic pen, a pencil, scissors, rubber, pottery, cooking foil and as many other things as you can find.

Make two piles, one of all the things that let the electricity flow and light up the bulb (remember that these are called conductors) and another pile of all the things that stop the electricity from flowing and don't light the bulb. These things are called insulators. What do you notice about all the conductors, what do they have in common?

A simple switch

It can be very useful to be able to switch a bulb on and off when you like without having to unscrew wires. If you trap a paper clip under one of the drawing pins in the circuit used to test conductors you can switch on the bulb by moving the other end of the paper clip to touch the second drawing pin. The paper clip is made from metal which conducts the electricity across the gap between the pins and completes the circuit, making the bulb light up. To turn it off again, simply move the paper clip away.

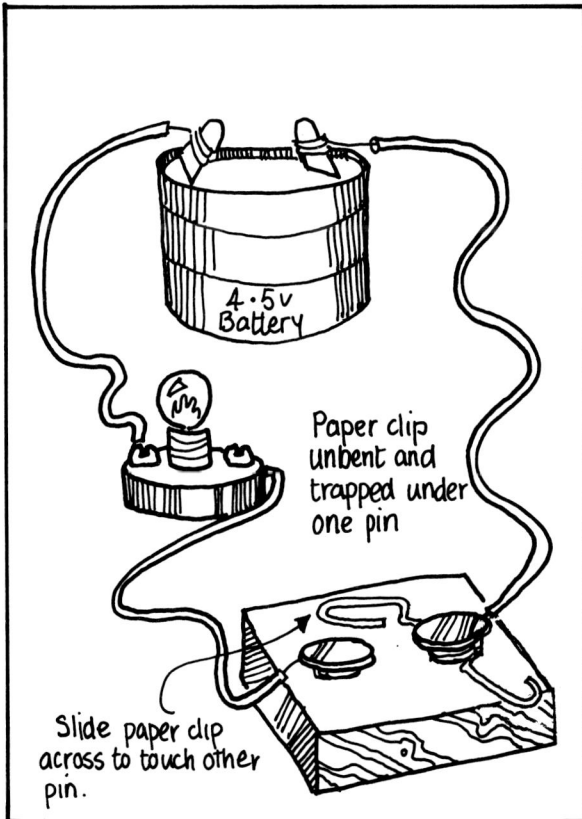

4·5v Battery

Paper clip unbent and trapped under one pin

Slide paper clip across to touch other pin.

A 'push to make' switch

wire to bulb holder

Push down
Paper clip bent up above 2nd pin

wire to battery

Another useful switch is called a 'push to make' switch. This is made in exactly the same way as the first one, except that the paper clip is bent up at an angle so that it is positioned above the second drawing pin. To turn on the light you have to push down on the paper clip. When you let go it springs back up again and the light goes out.

This kind of switch can be used to make a morse code key. You may be lucky enough to be able to buy a tiny buzzer which you can put in place of the bulb. Now you can send real coded messages to a friend, just give her a buzz!

Burglar alarm

Another interesting type of switch is a pressure switch. This works when something or someone treads or pushes down on the switch, setting off a light or buzzer.

It can be made very easily with a thin piece of card about 15 cm (6 inches) long by 8 cm (3 inches) wide, some cooking foil and a little paste or tape.

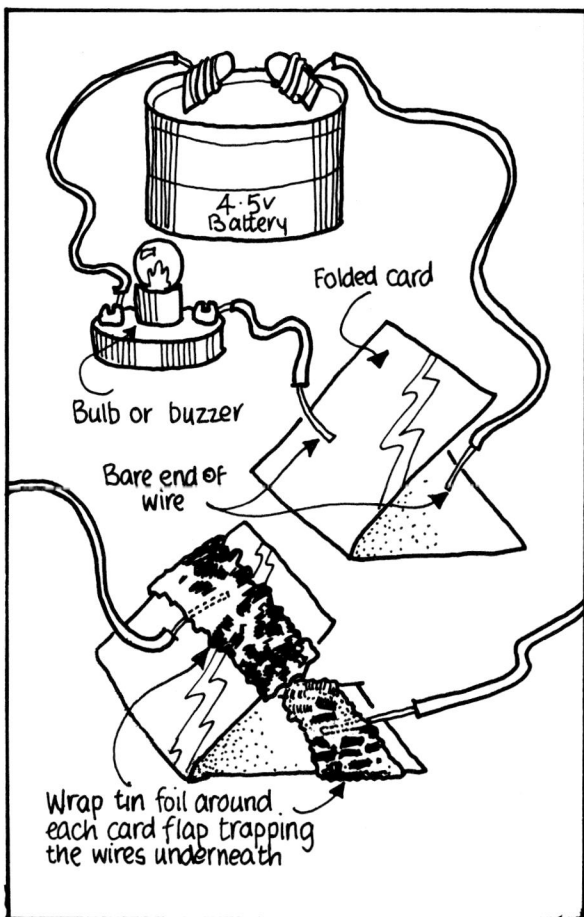

4.5v Battery

Folded card

Bulb or buzzer

Bare end of wire

Wrap tin foil around each card flap trapping the wires underneath

Make a simple circuit with a light bulb or buzzer like the one shown with the paper clip switch but, instead of trapping the wires under drawing pins pressed into a piece of wood, leave the ends of the wires free. Fold the piece of card in half and wrap strips of foil round the top and bottom flaps. Make sure that you trap one of the wires under each strip of foil.

The gap between the two strips of foil stops the circuit from being complete but, as soon as someone pushes down on the top flap of the card, the two strips of foil will touch and the bulb or buzzer will come on. For extra fun, position the switch near to a doorway so that any unsuspecting person walking in will tread on the switch and set off your alarm.

Magnetic attraction

People have known about magnetic force for hundreds of years. Sailors used chunks of magnetic rock called lodestone to help them navigate from one place to another. They had discovered that magnetic rock hung from a thread always pointed in the same direction, towards the North. Knowing where North was allowed them to set the right course for their home port.

Magnets are amazing, but a good one can be expensive and hard to find. Play magnets, sold in toy shops, are not very good for experiments. They are usually made of plastic with tiny pieces of magnetised metal glued to the ends. If you can find a supplier who will sell you a good magnet ask for a set made from cobalt steel. Look after them and they will last for years, but remember that any magnet that is dropped or banged will lose its magnetism.

Test your magnets on as many different

You can make your own by borrowing a magnet to magnetize a steel needle. As well as the magnet you will also need a slice of cork (cut from the end of a cork from a wine bottle), a dish, and a large steel needle such as a darning bobbin.

Magnetizing steel

Steel rods and needles can be magnetized by stroking them with another magnet. Lay the needle on a table top and, holding a magnet by one end, stroke the other end along the full

things as you can find. Try cups and saucers, glass jars and plastic bottles, aluminium pots and tin foil, keys, coins, rubber, brick and paper, knives and forks, wood, brass, paper clips and drawing pins.

Many things will not be attracted at all, but some things are attracted so well they will almost seem to jump at the magnet. Make a note of the things that are not attracted as well as those that are. What do you notice about the things that are attracted to magnets? What do they have in common?

Make your own magnetic compass

Lodestone gave sailors some idea of direction, but proper magnetic compasses are very accurate and are essential for navigating successfully across miles of open sea.

length of the needle at least twenty times. Always stroke the needle in the same direction and take the magnet well away before bringing it back to the needle's end to begin another stroke.

Setting up the compass

Half fill the dish with water and float the cork on the surface. Then balance the needle on the cork so that it floats straight and level in the water. Leave the needle to come to rest and you will find it will line up in the North – South direction.

Check to find which end is pointing North. (You can put a tiny blob of paint on the end to remind you). Wherever you are, if you set up your compass it will swing to point North.

Magnetism from electricity

Another way to magnetize an iron or steel rod is to put the rod inside a coil of wire. Electricity flowing through the wire causes the little magnetic particles, called domains, in the metal to line up and turn it into a magnet.

Try it for yourself, it is very easy. All you need is a 4.5 volt battery, some thin wire (the best kind to get is bare wire that has no plastic insulation around it) and a large nail about 15cm (6 inches) long.

Spend some time carefully and patiently winding the wire around the nail in a tight coil. Start 2 cm (1 inch) from one end of the nail and wrap the wire in neat coils around the nail until you get within 2 cm of the other end. You can keep the coils in place by wrapping a little Sellotape around them as you go.

Finally, connect the ends of the wire coil to the terminals of the battery. If you are using the bare wire it is quite likely that it has a coat of lacquer on it which must be cleaned off first. Rub 2 to 3 cm of each end of the wire with fine sandpaper to clean the ends and then connect up to the battery.

Unless you put a simple switch in the circuit, like the ones described in the electricity experiments, your electromagnet will come on straight away and will quickly waste the battery. Either add a switch or leave one wire disconnected until you are ready to use your electromagnet.

Get a box of paper clips or drawing pins.

Switch on the electro-magnet and dip the end of the nail into the box. The electricity flowing in the circuit magnetizes the nail which picks up the pins. You can try altering the number of turns of wire in the coil to see if it makes it weaker or stronger. Test the strength by counting the number of clips or pins that it picks up each time.

Light and sound

What is sound?

Touch your throat while you speak, can you feel the sound? Air from your lungs passes over your vocal cords making them shake or vibrate. These vibrations pass through the air and when they reach your ears the vibrations shake your ear drums and the vibrations are passed through tiny bones to nerves that run to the brain. Your brain makes sense of the patterns of vibrations and you recognise the sound as your own voice. All sounds are vibrations of one kind or another. Some sounds are very pleasing and musical while others we would just call noise.

Hold a ruler firmly on a table top with about half its length hanging over the edge. Twang the end down and the ruler vibrates with a funny sound. Now push the ruler further over the edge and twang again. The sound gets lower. Pull the ruler back so that only a few centimetres are overhanging and the sound is much higher. Grip one end of an elastic band firmly between your teeth and stretch it out. While you flick the band with one hand, tighten and loosen it with the other. You will hear a strange twanging noise which gets higher as you stretch the band and lower when it is relaxed.

A swaney whistle

If you have an old plastic recorder you can make some very interesting sounds. Cover all the finger holes with Sellotape. If you have done this properly it will play its lowest note when you blow gently into the mouth piece.

Find a large, deep jug or bowl and fill it with water. Place the end of the recorder into the water, take a deep breath and begin to blow gently. As you play dip the recorder deeper and deeper into the water. You will hear the note getting higher and higher. As the water gradually fills the recorder the air column inside gets shorter and the note gets higher. Move it up and down and you will hear a funny warbling note, just like a real swaney whistle.

Hearing through walls

Sound does not need to travel through air, it can pass through liquids like water and solid objects as well. When you are next in the bath lay down flat in the water and tilt your head back so that your ears go under the water. Lie still and listen. All sorts of strange sounds will come to you through the bath water, and they sound so much louder too.

Put your ear against a wall while your friend taps gently on the other side. You will be able to hear the tapping very clearly. Try the same thing on metal railings. Your friend can be some distance away but you will still hear her tapping.

The tumbler trick

If someone is talking in another room take a glass tumbler and place the open end right up against the door and put your ear tight against the bottom of the glass. The person talking is making the air in the room vibrate. These vibrations shake the door which makes the air in the glass shake too and you can often hear quite well what is being said.

Tin can telephone

Get two empty tin cans but make sure that they have no sharp edges. Use a hammer and a small nail to bang a hole in the bottom of both tins. Take about 3 metres of thin string and thread each end through the holes in the tins. Tie the end in a knot so that it cannot pull out again.

Ask a friend to take one of the cans and move apart so that the string is pulled tight. While you hold the can to your ear ask your friend to talk slowly and clearly into the other tin.

You should be able to hear clearly what she is saying. The vibrations from her voice shake the air in the tin and make the bottom of the tin shake too. These vibrations shake the string which passes the vibrations to the other tin which begins to shake as well. You hear the air, vibrating in your tin, as the sound of her voice.

How do records play?

Record players have a special kind of needle called a stylus in the end of the player's arm. When the stylus is placed into the groove of a record it vibrates. The record's grooves are lined with lots of lumps and bumps of different shapes and sizes. When the stylus hits these patterns of bumps it vibrates faster or slower, harder or more gently and sends electrical signals to the amplifier. We hear the signals as the sound of music and singing through the speakers.

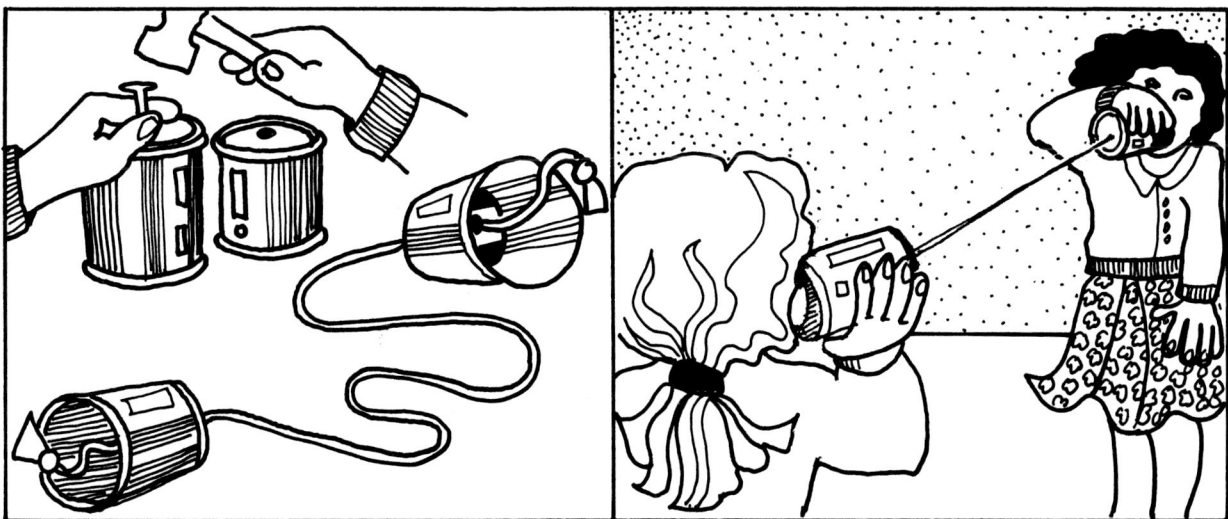

Yogurt pot record player

You do not need the record player arm and amplifier to be able to hear these sounds. Find a yogurt pot or margarine tub and push a pin through the base with a pair of pliers. Find a record that is old and no longer needed. Ask an adult to place the record on the turntable of a record player and set it turning at the correct speed. Hold your pot and place the needle into the groove of the record and listen with your ear close to the pot. You will be able to hear the sounds quite clearly.

The speed of sound

You may have heard of aircraft, like Concorde, passing through the sound barrier. This means that they travel so quickly they leave their own sound behind.

If you have been at a sports meeting and watched someone on the other side of the stadium starting a race with a gun, you may have noticed that you see the puff of smoke first and the noise of the bang arrives moments later. When watching a thunder storm the same thing happens. You see the bright flash of the lightning and the rumble of thunder follows a few seconds later.

This is because light travels so fast (186,000 miles in a second or 300,000 kilometres a second) that we see the flash instantly. The sound of the thunder moves much more slowly at about 760 miles an hour or about 330 metres every second.

To work out how far away a storm is, wait for a flash and begin to count slowly. For every 3 seconds the storm is 1 kilometre away (5 seconds for a mile). So a count of 9 seconds means the storm is 3 kilometres away.

Measuring the speed of sound

You can find the speed of sound by trying a very simple experiment. You will need a long tape measure, a metre stick or trundle wheel or some other method of measuring a distance of 500 metres. Ask a friend to take two tin lids or a drum or something very noisy and stand 500 metres away from you. You will need a stop watch that measures to the nearest tenth of a second. Many digital watches can do this. At a signal from you your friend should swing the tin lids or drum and drum stick together. Start

To work out the speed of sound you may need the help of a calculator. If you have five readings, add the times together and divide by five to get the average. Now double that time and that is how long sound takes to cover one kilometre. Divide the distance, 500 metres, by the average time in seconds and you will have calculated the speed of sound in metres per second. This is a very inaccurate method so don't expect to be even close to the correct figure of 330 metres per second. Anything between 200 and 400 would be a good answer. Perhaps you can think of a way of making your experiment more accurate.

Light and dark

We can see what is around us when light hits things and is reflected off into our eyes. Most of our light comes from the Sun, but we also get light from burning things, such as candles and fires, or a bulb that heats and glows when electricity passes through it, like the bulb in a torch or a lamp.

White light from the sun is actually made up of seven colours mixed together. We see these colours when sunlight is split by passing through raindrops. We call the colours a rainbow. When light falls on an object some of the colours are absorbed while others are reflected back. For example a ball that reflects back only blue light we will see as a blue ball. Things that reflect almost all the sunlight will look very light in colour, but those that absorb most of the light and reflect back very little will look black. That is why it is cooler in Summer to wear light coloured clothes, they reflect the light and heat of the sun. Black clothes absorb it and will make the person wearing them very hot.

timing the instant you see the things come together. Stop your watch again the instant you hear the noise. You will need to practice this, but once you are sure you can see and hear the signals to start and stop the watch, make a note of 4 or 5 sets of timings.

Splitting colours

Light is not the only thing that can be split into many colours. Many inks and dyes are mixtures of colours. If you have ever splashed some water on black ink you may have noticed that the ink smudges and runs leaving blue and brown trails over the paper.

This experiment is fun to do and fascinating to watch. You will need some strips of blotting paper about 20 cm long and 3 cm wide (8 inches by 1 inch), some felt tipped pens, a jam jar or similar sized pot, a pencil and some sticky tape. Put about 1 cm (half an inch) of water in the bottom of the jar. Tape a strip of blotting paper onto the pencil so that it hangs down from the middle. Make a big blob of colour about 3 cm (1 inch) from the bottom of the strip

with a dark coloured felt tipped pen. Use colours like dark green, brown or black. Rest the pencil on the mouth of the jar so that the strip hangs down inside and just touches the water.

Watch what happens very closely. You will see the water begin to soak into the blotting paper. It quickly rises up the paper and begins to smudge the ink blot. As the water continues to rise up the paper it takes the ink with it but you will soon notice something odd. The ink begins to separate out into other colours. When the water has moved over half way up the paper take it out and hang it up to dry. Try the experiment again with other colours. Some split into several different colours, but you will also notice that other colours do no split at all, they are pure colours, not mixtures. Try different kinds of inks and food dyes to see if they split as well.

You may have a problem with some special markers that are filled with permanent, water proof ink, these will not smudge or separate out, so only use water colours (the kind that wash off hands and clothes easily).

Coloured mushrooms

A different way to split colours is to cut a circle out of blotting paper about 15 cm (6 inches) across. Cut a narrow flap about 1 cm (half an inch) wide into the centre of the circle and bend it downwards. Put a large blob of ink (one that you know will split into lots of colours) at the very centre of the circle. Put some water in a cup and lay the circle on the cup so the flap dips down into the water.

Almost immediately the water will begin to soak up the flap of paper. When it reaches the ink blob it will spread the colours in rings. Take

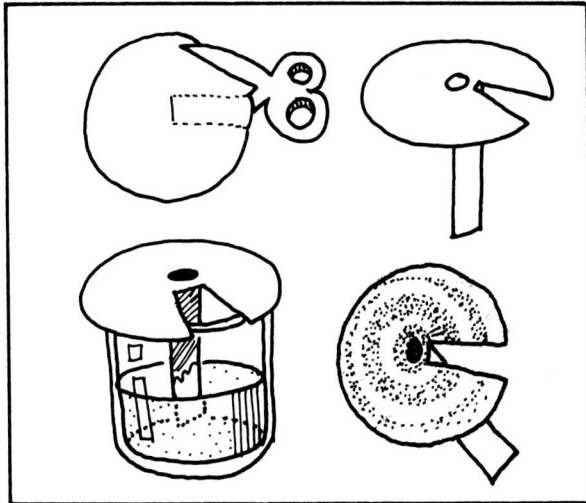

the circle off as soon as the water reaches the edge of the circle. When it is dry you will have a mushroom shaped piece of paper covered with rings of colour.

Shadow theatre

Light travels in straight lines. If it didn't, everything that we looked at would look out of focus and continuously moving with no sharp edges or definite shape. On a bright sunny day you can see the effect of light moving in a straight path from the sun. As the light hits you, you block its pathway. Because the light cannot curl around you there will be a dark shadow of yourself on the ground. Get a friend and see if you can make your shadows touch without actually touching each other. Can your shadows shake hands even though you are not actually holding your friend's hand?

Getting shadows to look as if they are doing things that are not actually happening can be great fun. Get an adult to help you set up a shadow theatre. All you need is a large cotton bed sheet and a bright lamp. Help to stretch the sheet across the centre of a room so

Broom stick with thin white cotton bed sheet fixed to it (with drawing pins perhaps)

Strong light source such as slide projector or powerful torch or a table lamp with shade removed or angle-poise type

Low table for operations!

SIDE VIEW

that you cannot be seen behind it. Place a low table behind the sheet and set up the lamp on a stool so that it shines towards the table and the sheet.

When it is dark ask your friends and family to come and sit on the other side of the sheet to watch a horrifying operation. Make sure that the only light that is on in the room is the lamp behind the sheet. When you walk about behind the sheet the audience will see only your shadow.

The operation

Lead one of your friends behind the sheet and lay them down on the table. (It is best to have arranged all this with them before the performance so they know what is going to happen. You will also need a few things to add to the effect like a large hammer, some rope, a big pair of pinchers and perhaps some cut out shapes to look like parts of the body). Say in a loud voice that you are going to have to

operate. Begin by explaining that you will have to put them out. Pick up the hammer and hit a pile of books that you have secretly placed next to your friends head (it will add to the illusion if your friend sits up, shouts "Ow!" and sinks back as if in a coma).

Take a large knife and pretend to cut open the body by slicing along their side. Pull up a length of rope and hold up all sorts of wierd shapes as though you are pulling them out of the body. Finally say that they are now cured and make it look as though you are piling the bits back in by putting them one by one onto the table beside your friend. Pretend to sew them up with a big cardboard cut out in the shape of a needle threaded with string. When you have finished the 'patient' miraculously comes back to life to the cheers of the audience.

It is always best to practice first so that you know what to say and what to do. Get someone who will keep your secret to watch your actions to make sure that the shadows can be seen clearly and the operation really looks as if it is happening. Only you and your friends will know that it is actually just a trick of the light

Bending light round corners

Although light travels in straight lines it can be made to bend round corners if you put the right thing in the way.

Have you ever looked into a swimming pool and thought how shallow it looks only to find that it is very deep when you have jumped or dived in. Water 'bends' light because it is thicker than air (we say water is more DENSE than air). Another way to see this happen is to hold a stick or a pencil in a large jar that is full of water. Look just over the lip of the jar into the water and it

Pencil held at a slight angle

Must look just over lip of jug or bowl

The pencil appears to bend over at an even greater angle

will look as if the pencil is bent at the point that it goes into the water.

Lenses and mirrors

Light can also be bent using lenses. If you are lucky enough to have a magnifying glass or a pair of binoculars you will already know that some lenses can make things appear much larger or nearer to you than they actually are. The lenses bend the light to focus it at a point.

59

Mirrors can also bend light. By reflecting the light they can change its direction. Shine a torch into a mirror in a darkened room. You can aim

your torch to light up things that are on the other side of the room or even behind you. The mirror reflects the light off at an angle, back into the room.

Up periscope

You can use mirrors to see over tall objects or round corners. Get two handbag mirrors, some card and a pair of scissors. The card should be about 40 cm (16 inches) square. Cut out the mirror slots and peep holes as shown, then fold the card into a tall box shape along the three fold lines. Tape the edges of the box together and wedge the mirrors into the slots. You can use your periscope to look over fences or round walls.

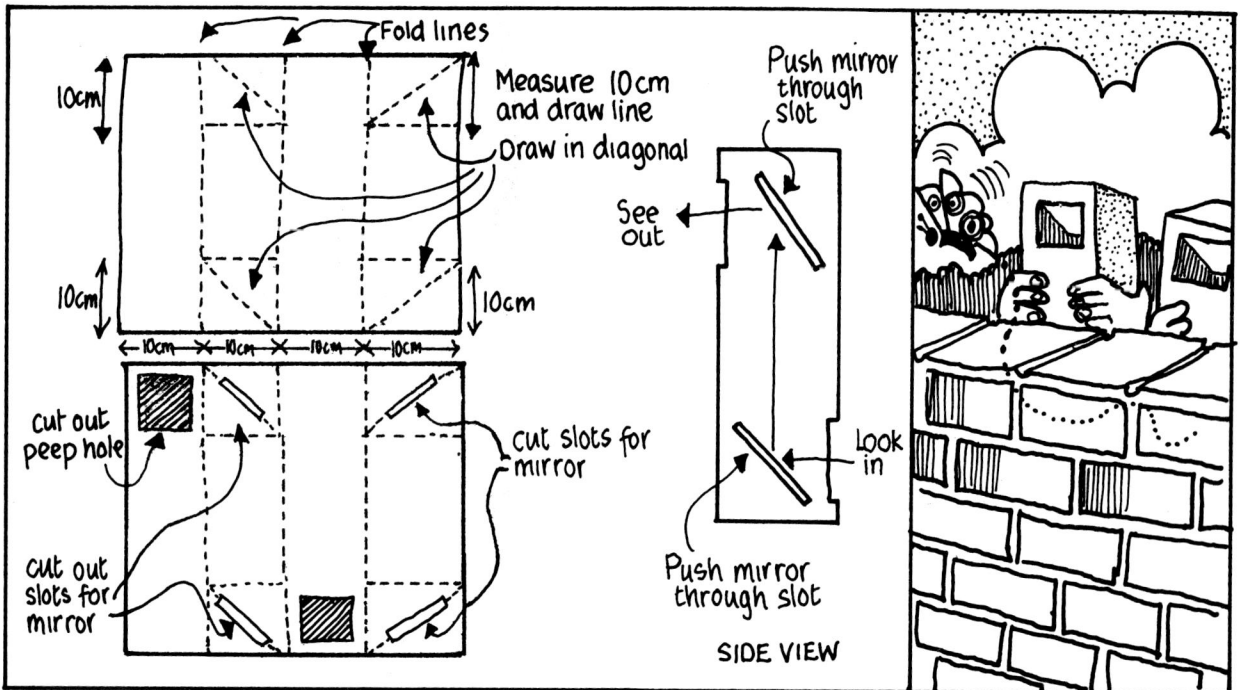

Marking time

Why are there 24 hours in a day? Why are there 365 days in a year? Why do we have night time and day time?

Time is very strange. It is something we take for granted but if we lived on another planet the length of a day or a year could be completely different. For example if we lived on Venus one day would last longer than one year!

The Earth is one of nine planets that circle around the Sun. The time that it takes a planet to go round the Sun is called a year. The Earth's year is just over 365 days. (We make up for the little time that we lose by having an extra day every fourth year, which we call a leap year). The planet Mercury is the closest planet to the Sun and has a much shorter path or orbit around the Sun than the Earth. Mercury's 'year' lasts only 88 of our days which is the time that it takes for Mercury to travel right round the Sun. Pluto on the other hand is the furthest planet from the Sun. Its journey around the Sun takes nearly 250 of our years.

The length of a day

All the planets spin as they move on their orbits around the Sun. So for part of the time a spot on the planet's surface is facing towards the Sun, which we call day time, and for part of the time it turns away from the sun and into darkness or night time. Since the times of the ancient Greeks and Romans the time that it takes the Earth to turn once has been called a day and the day was divided into 24 equal periods called an hour. But hours, minutes and seconds are man-made. There is no reason for having 60 minutes in an hour or 24 hours in a day, it was just decided upon. The Greeks

counted their day from one sunset to the next, for the Romans it was from midnight to midnight which we still use today.

The planet Venus turns so slowly that one 'day' lasts for 243 of Earth's days, but the giant planet Jupiter turns so quickly that its day lasts for less than 10 of our hours.

A simple shadow clock

Find a stick or a pole about one metre (3 feet) long. Knock it into a patch of dry, flat earth so that it stands straight up.

As the Earth turns around on its axis, for part of the time we are turned towards the Sun and for part of the time we are turned away from it. As we turn towards the Sun, the Sun appears to rise above the horizon towards the East (Sunrise) and, throughout the morning, it seems to climb higher in the sky until it reaches its highest point. Then during the afternoon the Sun appears to sink down towards the horizon, finally dipping out of sight over in the West (Sunset). The Sun is not actually moving, it is the Earth spinning round that makes it look as though it is.

You can watch what happens on a sunny day using your stick. Early in the morning look at the length and direction of the shadow of your stick on the ground. Use a sharp stone or a stick to scrape the exact length of the shadow into the ground. Do the same thing every hour, on the hour, throughout the day until evening. Look at the pattern of lines on the ground, what do you notice?

When the Sun looks higher in the sky, the stick's shadow gets shorter and shorter. It is at its shortest when the sun is at its highest point which is exactly midday. Look for your shortest line. What direction does it point in? Because

the Sun moves from the East to the West through the Southern part of the sky the shadow will point North. At exactly midday the stick's shadow will point directly towards the North.

Sundials

Look for sundials on the walls of old buildings or set on stone pedestals in the grounds. They use the Sun's shadow to help people find out the time. They are often shaped like a triangle or pointer leaning at an angle set into a metal plate that is marked with lines and numbers (often in Roman numerals). The sun casts a shadow of the pointer onto the base. If the shadow falls close to a line it will be near the hour shown by the number written on the line.

You can mark the scratches in the ground around your own shadow stick with the number of the hour when the line was made to make your sundial. But there is just one problem. In Britain, between the end of March and the end of October, clocks are put forward by one hour for British Summer Time. You will find that if you are trying to find the shortest shadow at this time of the year, the middle of the day when the sun is at its highest, will actually be at 1 o'clock according to our watches!

Time can be very confusing.

Phases of the moon

On a clear night at the time of a full Moon, the Moon appears so large and so bright you can almost read by its light. The light is actually coming from the Sun, reflecting off the surface of the Moon like a giant mirror. There are also other times when the Moon seems to have a bite out of it. Each night the bite gets larger and the crescent shaped Moon gets smaller and smaller until it is completely in shadow. This is called a new Moon. The shadow that covers the Moon is caused by the Earth being in the way of the Sun's light. From the time of a new Moon, as each day passes, the shadow gets less and less until, once again, the Earth is not blocking the Sun's light at all and the full Moon shines.

Keep a sketch book and pencil handy by your bedside. Remember to look out each night and try to spot the Moon. Some nights you will be unlucky because of thick cloud, but when you can see it, sketch the way that it looks. Try to show exactly how much can be seen and how much is in shadow. Put the date by each of your sketches. How many days does it take to go from full Moon to another? This period of time is called a Lunar month. A Lunar month (Lunar means something to do with the Moon) is always exactly the same length of time. Can you discover what it is?